Patterns of Progress

ON MICROBIAL PATHOGENICITY

Patterns of Progress

General Editor: J.Gordon Cook, B.Sc., Ph.D., F.R.I.C.

Concise, authoritative reviews of technical topics

Microbiology Series

Bacterial Conjugation, D.J. Finnegan
The Microflora of the Rumen, P.N. Hobson
The Mycobacteria, C. Ratledge
Scrapie in the Mouse, R.H. Kimberlin
Microbial Response to Mild Stress, R.E. Strange
Modern Views on Microbial Pathogenicity, H. Smith
Epidemiology and Infections, C.E. Gordon Smith
Yield Studies in Microorganisms, A.H. Stouthamer
Bacterial Reaction to Radiation, B.A. Bridges
Bacterial Transduction, S.B. Primrose
The Oxygen Metabolism of Microbes, D.E.F.Harrison
*Continuous Culture in Microbial Physiology and
    Ecology*, H. Veldkamp

MODERN VIEWS ON MICROBIAL PATHOGENICITY

Harry Smith, D.Sc., F.R.C.Path.
Professor of Microbiology, University of Birmingham
Birmingham, B15 2TT, U.K.

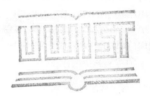

MEADOWFIELD

Meadowfield Press Ltd.,
I.S.A. Building, Shildon,
Co. Durham, England.

ISBN 0 904095 18 5

Printed in Great Britain at
Upton Printing Services,
Watford, England.

PREFACE

The Subject

This monograph summarises knowledge as at 1973 of the determinants of microbial pathogenicity and emphasises the gaps in this knowledge. Only broad principles can be covered in the space available but relevant examples are cited. Bacteria form the first and main subjects. Not only is more known about them than about other pathogenic microbes but effort in the bacterial field has revealed difficulties of investigation and produced broad concepts of pathogenesis which apply to studies of other microbes. The mechanisms of pathogenicity of viruses, mycoplasmas, fungi and protozoa are then discussed in the context of concepts used for bacterial pathogenicity. In each chapter a description of the main difficulties facing researchers in this field forms a prelude to a discussion of the following aspects of microbial pathogenicity:

(1) Entry to the host usually by surviving on and penetrating mucous membranes.
(2) Multiplication *in vivo*.
(3) Inhibition or non-stimulation of host defence mechanisms.
(4) Damaging the host.
(5) The reasons for tissue and host specificity.

CONTENTS

INTRODUCTION

The worst effects of infectious disease, especially death from bacterial disease, have been controlled in developed countries by public health measures. These measures were the practical consequences of proof that micro-organisms cause disease and the recognition of the relevant species by Pasteur, Koch and their associates in the 50 years prior to the first world war; and they have been achieved without sophisticated research on the mechanisms of infectious processes. Even vaccines and drugs, which played a significant but smaller part in controlling infection have been developed largely by empirical methods requiring little or no biochemical knowledge of the microbial or host determinants of pathogenicity. It is not surprising then that the nature of these determinants is still obscure except in a few areas of bacterial disease such as the classical toxaemias diphtheria, tetanus and botulism.

Why investigate the determinants of microbial pathogenicity if infectious disease is controlled? Despite the advances of the past century, infectious disease remains a major problem in human and veterinary medicine, with the economically important nuisance and chronic aspects gaining prominence as the fatal consequences become less frequent. Drug resistance of bacteria, protozoa and fungi is increasing. Effective chemotherapy of troublesome virus diseases is still lacking. Many vaccines remain unsatisfactory, either due to producing incomplete protection or to the hazard of injecting live organisms. The alarming increase in gonorrhoea is a current reminder of the ability of a once-regulated disease to rebound with changing social conditions. One of the most serious hospital problems is fatal septic shock following infection of debilitated patients by Gram-negative bacteria (*Pseudomonas aeruginosa, Klebsiella* spp.) that do not usually attack healthy people. Widespread dental caries and periodontal disease are other examples of the persistence of infectious disease. Obviously, new methods of attacking infectious disease are needed. Undoubtedly, as in the past, empirical screening procedures for drugs and vaccine strains will provide some new methods. However, after nearly 50 years, such procedures may be reaching the limit of their usefulness; so far they have failed to provide an effective therapy for virus diseases. There appears good reason for a more rational approach to the problem, namely that of attempting to recognise and then to neutralise the microbial and host determinants of pathogenicity. And this explains the re-awakening of interest in mechanisms of disease production.

Pathogenicity, Virulence, Markers and Determinants

The terms pathogenicity and virulence are nearly synonymous. They mean the capacity to produce disease. Comparisons of the pathogenicity or virulence of two organisms are made by determining the minimum number of each that are needed to produce a certain pathological effect in animals, e.g., those needed to kill half of a group of animals (the lethal dose fifty (L.D.$_{50}$)) or to produce a lesion of a certain size. Pathogenicity is used mainly with respect to species (e.g., *Bacillus anthracis* is more pathogenic than *Bacillus subtilis*) and virulence with respect to degrees of pathogenicity of strains within species (e.g., the Vollum strain of *B. anthracis* is more virulent than the Sterne strain). The classical method of investigating bacterial virulence has been to compare the properties of virulent and avirulent strains. The modern techniques of microbial genetics have increased the scope of the method by producing more strains. Studies *in vitro* on cell wall and capsular antigens, on metabolic characteristics, and on enzymes of different bacterial strains have indicated many virulence *markers*, i.e., factors associated with virulence. But it should be emphasised that these factors

(1) 2

are not virulence *determinants* until they have been shown to be produced *in vivo* during infection and to have biological activities directly connected with virulence, e.g., the power to inhibit phagocytes. Some products of virulent strains have been proved virulence determinants by their ability to promote pathogenic behaviour of an avirulent strain in infection experiments *in vivo* or in relevant tests *in vitro*.

This monograph summarises knowledge as at 1973 of the determinants of microbial pathogenicity and emphasises the gaps in this knowledge. Only broad principles can be covered in the space available, but relevant examples will be cited. Bacteria form the first and main subjects. Not only is more known about them than other pathogenic microbes, but effort in the bacterial field has revealed difficulties of investigation and produced broad concepts of pathogenesis which apply to studies of other microbes. The mechanisms of pathogenicity of viruses, mycoplasmas, fungi and protozoa are then discussed in the context of concepts used for bacterial pathogenicity. In each chapter a description of the main difficulties facing researchers in this field forms a prelude to a discussion of the following aspects of microbial pathogenicity.

(1) Entry to the host, usually by surviving on and penetrating mucous membranes.

(2) Multiplication *in vivo*.

(3) Inhibition or non-stimulation of host defence mechanisms.

(4) Damaging the host.

(5) The reasons for tissue and host specificity.

Most references for the work described will be found in the reviews and books that are cited and only very recent references not found in these sources are included in the text.

BACTERIAL PATHOGENICITY

Ten organisms of *B. anthracis* will kill a guinea pig whereas ten million of *B. subtilis* will not. Obviously pathogenic bacteria produce virulence determinants which enable them to invade a host and produce disease. The problem is to identify the number of determinants, their chemical nature, mode of action and order of importance.

Difficulties Of Studying Bacterial Pathogenicity

The fact that the pathogenicity of most bacteria such as staphylococci or streptococci is determined by a number of products and not by a single powerful toxin as in tetanus causes difficulty in recognising and fractionating any single determinant. But the main factor contributing to difficulty in this field is that virulence can be measured only *in vivo* and it is markedly influenced by changes in growth conditions due to selection of types and to phenotypic change. Bacterial virulence is at its maximum in bacteria obtained directly from infected animals. Usually it is reduced by sub-culture *in vitro* because under laboratory conditions bacteria lose the capacity to form one or more of the full complement of virulence determinants manifested in infected animals. Also, *in vitro*, apparent virulence determinants might be produced which are not formed *in vivo*, and therefore have no relevance. There is now abundant evidence for many species including streptococci, staphylococci, gonococci, anthrax bacilli and tubercle bacilli that organisms grown in infected animals are different chemically and biologically from those grown *in vitro*. Thus, although most studies are made on bacteria grown *in vitro*, these bacteria can be incomplete or misleading as regards the possession of virulence determinants. This is the essence of the difficulties encountered in studies of pathogenicity.

How then can the various factors involved in pathogenicity be identified? Obviously these factors can be produced in laboratory cultures if the correct nutritional conditions can be found. This has already happened for the classical bacterial toxins and some antiphagocytic substances. However, for problems of pathogenicity which have defied solution by conventional procedures using cultures *in vitro*, one approach would be to study bacterial behaviour *in vivo*. Aspects of pathogenicity might be revealed which later could be reproduced *in vitro* by appropriate changes in cultural conditions. This chapter describes several examples of virulence determinants that were recognised by gaining information on bacterial behaviour *in vivo* using one or more of the following methods. Bacteria and their products can be separated directly from the diseased host for biological examination and for chemical and serological study *in vitro*. The behaviour of organisms growing *in vivo* and their repercussion on the host can be examined either in the whole animal or in relevant tissues; a cardinal requirement for such studies is a suitable laboratory animal in which a natural infection of man can be simulated (e.g., the production of dental caries in germ-free rats by streptococci, or of cholera-like effects in rabbit gut by *Vibrio cholerae*). Some light can be shed on particular phases of microbial behaviour *in vivo* by making observations in tissue or organ culture and, finally, tests *in vitro* can often be made more relevant to microbial behaviour *in vivo*.

Investigating mechanisms of pathogenicity is difficult enough when only one type of organism is involved. But clinical disease is often due to mixed infection sometimes with organisms which are normally commensals. It is well known that one infection can have a profound effect on another, for example, the promoting effect of influenza on pneumococcal and staphylococcal infections. However, the difficulties of investigation are such

that we are only scratching the surface of the complex inter-
actions occurring in mixed infections. Comparative experiments
with single and mixed cultures must be conducted *in vitro* and
in vivo not only in normal animals but also in germ-free animals
where the experiments can be more defined. A few detailed
investigations of mixed infections have occurred and provide
templates for future work. In heel abscess of sheep a mixture
of *Corynebacterium pyogenes* and *Fusiformis necrophorus* is
needed for pathogenesis. The former provides a growth factor
and an anaerobic environment for the latter which produces an
aggressin (see p.10) that prevents phagocytosis of the former as
well as itself. Similarly, invasion of submucosal tissue in periodontal disease, which was simulated by subcutaneous infection
of guinea pigs, occurred only when the following four oral
organisms were mixed: *Bacteroides melaninogenicus*, another
bacteroides species, a motile Gram-negative rod and a faculta-
tively anaerobic diphtheroid. *B. melaninogenicus* was regarded
as the primary 'pathogen' and the diphtheroid produced for
it an essential growth factor, vitamin K. The roles of the other
two organisms have not been defined.

Entry: Survival On And Penetration Of Mucous Membranes

Although some bacteria enter the host directly by trauma or
vector bite, most infections start on the mucous membranes of
the respiratory, alimentary and urogenital tracts. Defence of
surface membranes and structures against microbial attack relies
on: (1) the mechanical flushing action of moving mucus or
lumen contents; (2) competition with and interference by surface commensals; and (3) the presence of bactericidal or
bacteriostatic materials in the mucous secretions. These defences
are overcome in infectious disease and electron and light micro-
scopy indicate at least three types of early attack on surfaces.
First, bacteria attach to the surface and multiply on it but do
not penetrate to any significant extent; this occurs in dental
caries, in respiratory infections with *Bordetella pertussis* and in
cholera and some *E. coli* enteric infections. Second, there can

be attachment and subsequent phagocytosis by mucosal cells with consequent surface damage as occurs in bacillary dysentery. Finally, there can be attachment and passage into the underlying tissues either through the mucosal cells or between them as occurs in streptococcal infections and in salmonellosis. The precise mechanisms underlying the three types of attack are not yet clear but there is much recent work on the problems.

In some cases we know the nature of materials produced by bacteria which stick them to surface membranes despite the flushing action of lumen contents, thus allowing them to proliferate at certain sites. Two dental caries bacteria, *Streptococcus mutans* and *Streptococcus sanguis*, produce from sucrose *in vivo* and *in vitro* a sticky dextran, the forerunner of dental plaque. Genetic transfer experiments with *E. coli* strains indicated the importance of the K88 protein antigen in attaching diarrhoea-producing strains to the brush border in the upper small intestine where they produce their enterotoxin. In addition to being antiphagocytic (see below) the M protein of *Streptococcus pyogenes* may help its attachment to epithelial surfaces in the mouth and throat. Gonococci appear to adhere strongly to urethral epithelial cells and possibly this is due to thread-like projections from the gonococcal surface called pili (Punsalang and Sawyer, 1973). Finally, observations on mutants of *Shigella flexneri* indicated 0 side chain structure (possibly rhamnose units) of the somatic antigen may be one determining factor for attachment and penetration of dysentery organisms into mucosal epithelial cells.

In trying to understand the complex ecology of commensals and pathogens on mucous membranes in relation to disease there are two essential questions to be answered. First, how do small inocula of extraneous pathogens such as the cholera vibrio survive and grow in competition with normal commensals? Second, in what circumstances and by what mechanisms do normal commensals adopt a pathogenic role? The second question is particularly relevant to diseases like dental caries. Unfortunately, in most cases the answer to both questions is that we do not know, although there have been some interesting peripheral studies. Qualitative and quantitative assessments of the different members of bacterial populations on different mucous surfaces in animals of different ages, in animals on different diets and in animals treated with antibiotics have shown that the normal flora varies with site, age and diet and that one organism can be antagonistic to another. For example, the removal of bacteria, especially lactobacilli, from the alimentary tracts of animals with penicillin usually results in an increase of *Candida* infections. The biochemical bases for these changes in flora are largely unknown. Population experiments have emphasised that the environment on some mucous surfaces is anaerobic since only anaerobes survive as commensals; they have thus underlined the difficulty that aerobic pathogens may have in initiating infection in these areas. Commensal competition with some extraneous pathogens has been investigated *in vitro* and in germ-free and normal animals. The nature of some bacteriostatic materials is known; for example, fatty acids produced by intestinal fusiform bacteria are inhibitory, in a reducing environment, to typhoid and dysentery organisms. But we still do not know how these inhibitory acids are overcome in the initial stages of dysentery and typhoid. Concentration of potentially pathogenic commensals in one place by dietary influences may start pathogenic processes, for example, the influence of a sucrose diet on dextran production by dental plaque organisms. Also, as discussed previously, the correct mixture of otherwise non-pathogenic commensals may be needed in some cases.

The manner in which pathogenic bacteria inhibit the third defence mechanism of surfaces, the bactericidal or bacteriostatic materials in the mucous secretions, will probably be similar to that discussed below (p.10) for humoral bactericidins in serum.

Multiplication In Vivo

Virulent bacteria must multiply in the host tissues in order to produce their disease syndrome. Two qualities are needed for multiplication. First, an inherent ability to multiply in the physical and biochemical conditions of the host tissue and, second, an ability to inactivate or not to stimulate host defensive mechanisms which would otherwise kill or remove them. The effects of these two qualities *in vivo* are not easy to separate and often it is difficult to assess their relative importance in the overall proliferation of the invading microbe. In this section the first quality − ability to multiply − is discussed.

Avirulence can arise from inability to grow and divide in the environment *in vivo*. Thus, nutritionally deficient mutants of *Salmonella typhi* were avirulent unless injected with their required nutrients. However, for most bacteria the tissues and body fluids probably contain sufficient nutrients to support some growth. Few naturally-occurring strains will be avirulent due solely to inability to grow in the host. Nutritional considerations will, however, affect rate of growth *in vivo*. The more rapid it is, the greater the chance of establishing the infection against the activity of the host defence mechanisms. What do we know of multiplication rate *in vivo*? The numbers of viable bacteria in the tissues of an infected host can be counted at any time after inoculation. But these numbers are only the resultants of multiplication and destruction or removal and only recently has a method been evolved for measuring true bacterial division rates *in vivo*. Pathogenic bacteria, genetically labelled with markers retained by half the progeny at each division were injected. The division rate was calculated after measuring the proportion of organisms with the marker at various times after inoculation into animals. Remarkable results were obtained. In the spleens of mice, *Salmonella typhimurium* divided at only 5-10% of the maximum rate *in vitro*. This type of approach may be used in the future for measuring true division rate of other microbes *in vivo*.

We shall see later how variation of the biochemical conditions for bacterial multiplication can explain some examples of tissue preference of pathogenic bacteria.

Inhibition Of Host Defence Mechanisms

To increase within the host tissues, metabolic ability to multiply in the nutritional environment is not enough. Pathogenic bacteria must also be able to inhibit host defence mechanisms which otherwise would destroy them. The bacterial compounds which inhibit these mechanisms are called 'aggressins', an old term which well describes their biological role.

Aggressins act in the decisive, primary lodgement period of infection; that is, during the first few hours when the few invading bacteria are most vulnerable to the protective reactions of the host. At this early stage, aggressins must inhibit non-specific bactericidal mechanisms; not only those already existing in or on the tissues but also those agencies, especially phagocytic cells, that are mobilised by inflammatory processes soon after the tissues are irritated. If some bacteria survive the primary lodgement and grow, spread of infection is opposed by the fixed phagocytes of the reticulo-endothelial system (lymph nodes, spleen, liver); and again, to make headway, bacteria need aggressins, possibly different from those operating during the early lodgement phase. To break through the protection of previously immunised animals or of animals several days after infection, bacteria must either be numerous or well endowed with aggressins, since the host defence mechanisms are of increased efficiency and are supplemented by antibodies capable of direct neutralisation of microbial products. It is possible that some aggressins inhibit the onset of the immune response. The clinical outcome of the disease depends on the interplay of these defensive reactions of the bacteria and of the host, and varies from complete subjugation of the host to complete destruction of the bacteria, and includes near stalemate in chronic infections and carrier states. Various types of bacterial aggressins are described below.

1. *Inhibitors Of Blood And Tissue Bactericidins*

Body fluids (blood, saliva, mucus) and tissues contain a variety of bactericidal factors such as basic polypeptides, lysozyme, complement (acting with antibody or non-specific substances), and possibly a system involving the iron-binding protein transferrin. Since clearly there are several different types of bactericidins, virulent bacteria must produce different types of aggressins to inhibit them. Resistance to these bactericidins has been associated with virulence in strains of many bacterial species such as streptococci, enteric pathogens, staphylococci, anthrax bacilli, brucellosis bacilli and gonococci isolated directly from patients and tested without subculture. But only rarely have the aggressins been chemically identified. Those from anthrax bacilli are capsular poly-D-glutamic acid and the three-component anthrax toxin (see later). Resistance of *Brucella abortus* to the bactericidal action of bovine serum is due to a cell-wall component containing protein, carbohydrate, formyl residues and much (35-42 per cent) lipid. The acid polysaccharide K antigens of *E. coli* from urinary infections seem to inhibit the action of serum and complement.

2. *Inhibitors Of The Action Of Phagocytes*

Once a microbe has penetrated into the tissues, phagocytic activity of the wandering and fixed cells of the reticulo-endothelial system forms the main protective mechanism of the body: a mechanism which acts non-specifically but which is greatly enhanced by immunisation. Phagocytes vary in origin, morphology, constituents and bactericidal function. There are two main types each having two subdivisions: polymorphonuclear (neutrophils and eosinophils) and mononuclear (blood monocytes and tissue macrophages) phagocytes. Polymorphonuclear phagocytes are end cells with a short life derived from different stem cells from the long-lived mononuclear phagocytes. Inflammatory exudates contain cells of all types, the polymorphonuclear cells predominating initially but later dying to leave the mononuclear phagocytes ascendent. Macrophages form the fixed phagocytic system in the lymph-nodes, spleen and liver.

Bacterial products can interfere with the action of phagocytes by preventing one or more of the following processes, phagocytic mobilisation (inflammation), contact with bacteria, ingestion and intracellular killing.

Inflammation

In mice, virulent staphylococci suppress the inflammatory response, multiply rapidly and produce severe lesions. This anti-inflammatory activity is due to a cellwall component, probably a mucopeptide, but its significance in human infections is not known.

Contact

Contact with bacteria is effected by random hits, by trapping in the filtration systems of lymph-nodes, spleen and liver, and by chemotaxis. Bacterial products could hardly interfere with the mechanical processes, but fractions from tubercle bacilli and staphylococci interfere with chemotaxis *in vitro*. However, the significance of these observations in infection is unknown.

Ingestion

Once engulfed by phagocytes, many bacteria (salmonellae, pneumococci, staphylococci, streptococci, anthrax and plague bacilli) are usually destroyed and digested by discharge of the phagocytic granules into the vacuoles surrounding the ingested bacteria. Resistance to ingestion, thus avoiding intracellular destruction, is the main aggressive mechanism of these bacteria. There are two types of aggressins: surface and capsular products which do not appear to harm the phagocytes, and toxic materials producing direct damage. Examples of the first are the capsular polysaccharides of the pneumococci and meningococci, the capsular hyaluronic acid of streptococci, the 0 somatic antigens of some entero-bacteriacae, the capsular polyglutamic acid of *B. anthracis*, the Vi antigen (poly-N-acetyl-D-galacto-samino-uronic acid) of *Salm. typhi*, the acid polysaccharide K antigens of *E. coli*, the protein carbohydrate envelope substance of *Pasteurella pestis*, a cell-wall mucopeptide from staphylococci and a surface slime from *Ps. aeruginosa*. Examples of the second are the leucocidins of the staphylococci, the anthrax toxic complex and the M protein of streptococci.

Investigations of the relation between structure and anti-phagocytic activity have been carried out using knowledge of the chemical structure of the 0 antigens of mutants of the enterobacteriacae. Resistance of *E. coli* to phagocytosis by mouse polymorphonuclear leucocytes appears to depend on a complete saccharide component in the 0 antigen; a mutant lacking only colitose in its side chain was significantly more phagocytosis-susceptible (and less virulent) than the wild type and a mutant lacking galactose, glucose, N-acetylglucosamine and colitose was even more so. Similarly, full phagocytosis resistance and virulence of *Salmonella typhi-murium* for mice seems to depend on the tetrasaccharide sequences abesquosyl-mannosylrhamnosyl-galactose in the 0 antigen, acetyl and glucosyl groups being less important.

The mode of action of these aggressins is still not clear. Non-toxic aggressins may interfere with ingestion by purely mech-anical means, by inhibiting the adsorption of serum opsonin as seems to occur for *B. anthracis* and staphylococci, and possibly by rendering the bacterial surface less foreign to the host. Toxic aggressins probably harm phagocytes in the same way as they affect ordinary cells; for example, interference with membrane function by the leucocidin of the staphylococci.

Digestion

Virulent strains of some bacteria resist the phagocytic bacteri-cidins which destroy other bacteria and thus they survive or grow intracellularly. Within the cells, the bacteria are protected from natural and injected anti-bacterial agents and hence they often produce chronic diseases. Tubercle bacilli, leprosy bacilli, brucellae and *Listeria* spp. are the typical 'intracellular' pathogens; ability to grow within phagocytic and other cells is probably the most important aspect of their pathogenicity and can be seen both in infected animals and in cell maintenance culture *in vitro*. In addition, bacteria whose virulence is deter-mined in part by resistance to phagocytic ingestion can, when ingested under certain circumstances, survive and grow intra-cellularly. Examples are salmonellae, shigellae, plague bacilli and staphylococci.

Although much has been learned of the bactericidal mechanisms of polymorphonuclear and mononuclear phagocytes, practically nothing is known about what happens to bacteria when they are killed by these mechanisms. And studies of how intracellular pathogens resist the phagocytic bactericidins have only just begun.

Virulent strains of *Br. abortus* survived and grew in bovine phagocytes more than avirulent strains. This was neither due to a greater ability of the virulent strains to use the nutritional conditions within the phagocytes nor to their higher catalase content which may have afforded a greater protection against the bactericidal action of phagocytic hydrogen peroxide. It appears to be due to production under the growth conditions occurring *in vivo*, and simulant ones *in vitro*, of a cell-wall substance which interferes with the bactericidal mechanisms of phagocytes. Virulent brucellae obtained from infected bovine placental tissue or from cultures in laboratory media supplemented by bovine placental extract or foetal fluids had an increased ability to survive intracellularly compared with the same strain grown in laboratory media. Cell-wall preparations of the organisms from infected bovine placenta and from the supplemented media, inhibited intracellular destruction of an avirulent strain of *Br. abortus*. Finally, from *Br. abortus* grown in supplemented medium, the antigen which appeared responsible for inhibition of the phagocytic bactericidins was removed by washing with an ether-water mixture.

Mycobacterium tuberculosis and *Mycobacterium microti* (causal agent of vole tuberculosis) appear to resist intracellular bactericidins by a different mechanism from *Mycobacterium lepraemurium* (mouse leprosy). In mouse peritoneal macrophages infected with *M. tuberculosis* and *M. microti* the lysosomes (phagocytic granules) do not discharge into the phagosomes (phagocytic vacuoles). On the other hand, *M. lepraemurium* in mouse peritoneal macrophages and in rat fibroblasts survives and grows despite lysosomal discharge into the phagosomes. The nature of the aggressins determining the different types of resistance is unknown. Possibly they are connected with the electron-transparent layers, apparently containing wax fibres, that surround these organisms *in vivo*.

3. *Inhibition Of Immune Responses*

Surprisingly, more work has been done on immunosuppression by viruses than by bacteria, although this might be an important aggressive effect of bacteria. Group A streptococci and *E. coli* contain cytoplasmic factors which inhibit antibody response to sheep red blood cells. L-asparaginase of *E. coli* prevents the appearance of plaque-forming cells making antibody to sheep blood cells. Cell-mediated immunity seems to be depressed in burned patients infected with *Ps. aeruginosa* and in tuberculosis, whooping cough and lepromatous leprosy. The bacterial compounds responsible for these effects are unknown but for Gram-negative infections endotoxin may play a role since it is known to decrease antibody formation.

Damaging The Host

When a bacterial species produces a fatal or serious disease two experimental situations arise with regard to the products it forms *in vitro*. First, poisons or toxins are produced *in vitro* or, second, they are not. These situations are discussed in turn.

1. *Bacterial Toxins Are Produced*

The toxic activities of bacteria can be divided into four categories.

Toxins responsible for non-infectious disease because they are produced outside the host
Bacteria can produce poisons in foodstuffs; the toxin of *Clostridium botulinum* and the enterotoxin of staphylococci are the main examples. The disease that occurs on ingestion of the infected food material is a chemical poisoning comparable

to that caused by eating a poisonous plant, fungus or shell-fish.
It is not an infectious process, but clearly the microbial toxin
is responsible for disease. Botulinum toxin has the general
characteristics outlined in the next section; it is a protein
neurotoxin acting on the autonomic system by interfering with
acetylcholine synthesis or release. The staphylococcal enterotoxin is the one microbial toxin for which a full aminoacid
sequence is available.

Toxins of overriding importance in infectious disease
Clostridium tetani and *Corynebacterium diphtheriae* produce
in vitro powerful exotoxins that have been well characterised
and are responsible for almost the whole disease syndromes,
since immunisation with toxoid (formalin-treated, detoxified
toxin) protects against disease. Both toxins are proteins with
no toxic moieties or abnormal aminoacids. They are the most
poisonous materials known to man. Tetanus toxin is a neurotoxin acting on the central nervous system, possibly by interfering with synaptic inhibitors. Diphtheria toxin interfers with
protein synthesis and consists of two parts, one determining
entry into a cell and the other inhibition of a transferase which
halts protein synthesis (Uchida, Pappenheimer and Harper,
1972).

Toxins which are significant but not the only factors respon-
sible for infectious disease
These toxins were recognised originally in cultures *in vitro* and
can be responsible for some pathological effects of infection.
However, they are not the sole determinants of disease, for
often as much toxin is produced by avirulent as by virulent
strains, sometimes injection of toxin does not produce all the
pathological effects of disease and usually immunisation with
toxoid does not confer solid protection against infection.
Examples are the α-toxin of staphylococci, which alters membrane permeability of many types of cells, and the rash-forming
toxin of streptococci. But the most important representatives
of these toxins are the endotoxins, lipopolysaccharides inti-

(1) 16

matcly associated with the cell-walls of many different Gram-negative bacteria. The basic structure consists of a heptose polymer core with lipid moieties containing 2-keto-3-deoxy-octanoic acid and polysaccharide side chains attached. When extracted from cell-walls by fairly drastic means (treatment with trichloracetic acid or warm aqueous phenol) and injected into animals they produce toxic manifestations – pyrexia, diarrhoea, prostration and death. In some infections, endo-toxins are liberated from the cell-walls of invading bacteria and are responsible for pathological effects, such as pyrexia, shock and death in typhoid fever or, pyrexia and shock in brucellosis of man, and abortion in brucellosis of domestic animals. On the other hand, we shall see later that in some Gram-negative infec-tions the pathological effects are due to an exotoxin and not to the cell-wall endotoxin, which is never liberated in significant quantities.

Toxin produced in vitro but of unknown importance in disease
Many substances, producing toxic effects related or unrelated to disease syndromes, have been isolated from cultures. Some may be laboratory artifacts having no relevance to disease *in vivo*. Even if formed *in vivo*, they may play no significant role in infection. Examples are some of the many enzymic products of staphylococci, streptococci and other organisms.

2. *A Serious Disease In Vivo And A Toxin Has Not Been Produced In Vitro*

This situation still occurs in bacteriology and it is the general case for diseases produced by microbes other than bacteria. There are two explanations. First, a toxin is produced but has defied recognition; second, host damage is effected by processes other than direct toxicity of microbial products.

Hitherto unknown toxins recognised by studying bacterial behaviour in vivo or in biological tests relevant to the disease
Several important bacterial toxins have been recognised in this way. The first was a toxic complex now generally accepted as

the cause of death in anthrax; it was recognised in the plasma of guinea pigs dying of anthrax. Later it was reproduced *in vitro*, purified and shown to consist of three synergistically acting components, two proteins and the other a metal-chelating agent containing protein, some carbohydrate and phosphorus. In the past decade the role of toxins in acute diarrhoeal diseases of man and animals has become clear by discarding mouse toxicity tests in favour of more relevant biological and animal tests in which organisms and their products are examined for their effects in the gut. Investigations on cholera formed the template for those on other diseases.

An enterotoxin from *V. cholerae*, responsible for the gross fatal fluid loss from the intestine which occurs in cholera, was recognised by using two tests. First, a ligated segment of small intestine in a living rabbit would fill with fluid following intra-luminal injection of *V. cholerae* and its products. Second, *V. cholerae* and its products caused fluid accumulation and diarrhoea in suckling rabbits when introduced into the gut lumen by a gastric tube. The extracellular enterotoxin has been purified and its mode of action studied. It is a heat labile protein with a molecular weight of 90,000 and it is different from the cell-wall endotoxin. It acts by increasing normal secre-tion of the small intestine, possibly by activating adenyl cyclase present in the intestinal epithelial membrane, thereby raising intracellular cyclic adenosine monophosphate levels, which in turn would affect electrolyte transport. Simple replacement of fluid and electrolytes usually leads to rapid and complete recovery from the disease.

Using similar 'gut reaction' tests, enterotoxins have now been demonstrated for the following diarrhoea-producing organisms: *E. coli* (scours in young pigs and calves and diarrhoea in babies; enterotoxin production is plasmid transmitted and the entero-toxin activates adenyl cyclase), *Vibrio parahaemolyticus* (a marine vibrio which in the summer causes the majority of the food poisoning cases in Japan) and *Clostridium perfringens* (food poisoning in man). *Shigella dysenteriae* also produces a similar toxin but mucosal invasion seems to be more important in the pathology of dysentery than enterotoxin production.

These recent discoveries of relevant toxins in several important diseases are a warning against attributing damage in other microbial diseases to causes other than direct toxicity until the possible production of toxins has been thoroughly investigated using realistic biological tests.

The role of immunopathology in bacterial disease
Anyone who has suffered hay fever or asthma will know that the immunological reactions of the host, although usually protective, can sometimes have unpleasant consequences. Classical work with *M. tuberculosis* in guinea pigs has shown that hypersensitivity to bacterial products can be dangerous and even fatal for the host. Furthermore, skin tests indicate that hypersensitive states occur in many bacterial diseases such as streptococcal infections, staphylococcal infections, pneumococcal infections, brucellosis, tularaemia, glanders, leprosy and salmonellosis. The hypersensitive reactions are usually of the delayed type, indicating that cellular mechanisms are involved, but antibody mediated reactions can also occur, e.g., against bacterial polysaccharides. Thus, potentially in many diseases, non-toxic bacterial products could produce harm by evoking hypersensitivity reactions. But, just as production of a toxin *in vitro* does not necessarily mean that it is relevant *in vivo*, mere demonstration of a state of hypersensitivity by a skin test is no proof of the implication of hypersensitivity reactions in the main pathological effects of the disease. More extensive investigations are needed; the main systemic and local effects of the disease must be simulated by hypersensitivity reactions evoked in a sensitised host by products of the appropriate microbe. In some cases this has been done. There seems little doubt now that the pathology of tuberculosis is largely due to hypersensitivity to products, particularly the waxes, of *M. tuberculosis*. Hypersensitivity also appears to play a role in cardiac and kidney lesions following infection with strepto-

cocci, although acute effects seem to be due to direct toxicity of streptococcal products on the susceptible tissues. Delayed-type hypersensitivity also appears important in host damage in leprosy, chronic brucellosis and syphilis.

Tissue And Host Specificity In Bacterial Infections

Why, in man, do *Streptococcus mutans*, diphtheria bacilli, pneumococci and meningococci show predilections for the teeth, throat, lung and meninges respectively? Why is gonorrhoea confined to man and Johne's disease to cattle and related species? The two most likely explanations for differences in susceptibility to infection between different tissues or hosts, are differential distributions of bactericidal mechanisms and differential distributions of nutrients for which the metabolism of the parasite is especially adapted. Despite much effort, attempts to lay the responsibility for specificies of single infections unequivocally on variation of defined bactericidal mechanisms have so far failed. However, there is some evidence that kidney tissue is prone to a number of infections due to inhibition of complement and of phagocytosis by the high pH and salt concentrations respectively in this site.

More success has been achieved in investigations of the influence of nutrition. *Corynebacterium renale* and *Proteus mirabilis* persist in and cause severe damage to the kidney of cattle and man respectively. These localisations appear to be due to possession of ureases which enable the bacteria to use urea for growth and for production of ammonia which damages the tissue. Brucellosis in many animals (e.g., humans, rats, guinea pigs and rabbits) is a relatively mild and chronic disease; the causative organisms do not grow prolifically and have no marked affinity for particular tissues. However, in pregnant cows, sheep, goats and sows there is enormous growth of brucellae in the placentae, foetal fluids and chorions, leading to the characteristic climax of the disease — abortion. Investigations have shown that the presence of erythritol, a growth stimulant for brucellae, in the susceptible tissues of susceptible

(1) 20

species explains this tissue specificity in brucellosis. Thus, in cattle, the foetal placentae, the chorion and foetal fluids contained more erythritol than the other foetal tissues and the maternal tissues contained no erythritol. Furthermore, the foetal placentae of cattle, goats, sheep and sows contained erythritol, but not those of man, rats, guinea pigs and rabbits. Finally, the S19 vaccine strain of *Br. abortus* which has been used safely without significant abortion in the field was inhibited by erythritol. *Strep. mutans* and *Strep. sanguis* localise in dental plaque because of a nutritional influence, that of sucrose in the diet. However, the mechanism is not a direct stimulation of growth so much as the production of dextran which sticks the organisms on the teeth and possibly on damaged heart valves in endocarditis.

An example of host resistance being determined by nutritional influences involves Brazilian strains of the plague bacillus. These did not kill guinea pigs which are killed by normal strains because, unlike the latter, they required asparagine to grow well and guinea pig serum contains a powerful asparaginase.

VIRUS PATHOGENICITY

Viruses differ from other microbes in having a unique method of replication involving an intimate association with the metabolism of the host cell. But viral, like bacterial, pathogenicity is not determined solely by biochemical ability to replicate in the host tissues. Virulent and attenuated strains replicate in host cells *in vitro*, yet they differ fundamentally in behaviour *in vivo*, presumably − as for bacteria − due to different capacities to counteract host defence mechanisms and to damage tissues. Also, the fact that virus factors responsible for virulence mechanisms are induced within host cells does not confer uniqueness on viral pathogenicity; although replicating by different processes, many pathogenic microbes including bacteria are intracellular parasites producing their virulence factors intracellularly.

Difficulties Of Studying Viral Pathogenicity

Studies on mechanisms of viral pathogenicity suffer from the same difficulties as in the bacterial field; namely, that pathogenicity is determined by more than one factor and can only be studied *in vivo*, where viruses often behave differently than in the more conveniently studied tissue culture systems. In addition, the first essential for studying the subject, quantitative comparison of the virulence of different strains, is inaccurate. Disease effects in animals (LD_{50}; lesion size; or mean death time for the 'same' dose) must be related to amounts of virus particles indicated by plaque counts or egg infection. The latter detect only a small proportion of the total virus particles and therefore may not measure all the particles (which could vary for different strains) capable of multiplying in experimental animals. For example, plaque counts on chick embryo fibroblasts detected less infectious particles of Semliki Forest virus than infection of suckling mice and in this system the proportion of total virus particles detected by the plaque counts was fairly high (approximately 1 in 10) compared with many other virus systems. Thus, for this and other reasons, comparisons of the virulence of virus strains are often imprecise. Hence only virus strains for which conventional tests have indicated large differences in virulence should be compared to recognise virulence markers and determinants. Comparisons of such well-tested and well-separated strains have been rare but informative, and they indicate the potential for the future use in the virus field of the classical method of bacteriology. Attenuated strains of poliovirus had less affinity than virulent strains for the receptors of cells of primate central nervous system. Virulent strains of ectromelia virus had a greater capacity to infect mouse macrophages than attenuated strains, although they had equal ability to infect hepatic cells. The virulence of Newcastle Disease virus strains was associated with an increased capacity to produce cell fusion effects and plaques in chick embryo fibroblasts; there were no major differences between strains in the kinetics of replication, either in timing or amount of virus released.

Entry: Survival On And Penetration Of Mucous Membranes

Detailed knowledge of the factors influencing mucosal invasion of viruses is lacking, due to the absence of techniques for observing the behaviour of a few highly dispersed virus particles on mucous surfaces with their indigenous microbial populations. As for bacteria, the main mucosal defences against viruses are inhibitory materials in mucus, the flushing action of moving mucus and lumen contents and probably the competitive action of commensals. The mechanisms whereby viruses overcome these defences are unknown. Early virus attack of the respiratory tract has received some attention and the site of membrane lesions and the preferentially-attacked cell types have been observed in Newcastle Disease of chickens and influenza of experimental animals and man. Nevertheless, there appear to be few deeper investigations on interference with mucus inhibitors, adherence to the mucosal surface and overcoming influence of commensals; Nor on the mechanisms of entry of various viruses into the particular epithelial cells they select for attack. A comparison of the resistance of virulent and avirulent strains to mucus inhibitors might show that resistance correlated with virulence and then lead to identification of a virion component responsible for the effect. An affinity for the surface receptors or the phagocytic action of susceptible mucosal cells against which virus particles brush may result in adherence of the virus to the mucosal surface and thus counteract the mechanical flushing action of mucus and lumen content movement. For example, influenza virus adheres very strongly to the cilia and surfaces of ferret nasal mucosal cells, but the virus and host components concerned are unknown.

Replication In Vivo

Although ability to replicate in host tissues is not the only factor in viral virulence, it is essential, and the more rapid the rate of replication the more likely the success of the virus in producing its disease syndrome.

Investigations of the biochemistry of virus replication in animals are complicated by the obligate parasitism involved, which not only entails complexity of the factors required for replication but also increases the difficulty of distinguishing the influence of their absence from that of host factors (defence mechanisms) which actually destroy virus or inhibit replication. The ability of a virus to replicate in a particular cell depends on inherent features of that cell. These features can be involved in one or more stages of replication: attachment and penetration of virus, uncoating, provision of energy and precursors of low molecular weight, synthesis of viral nucleic acid and proteins, assembly and release. The characteristics of the host cell which determine these stages of replication might be called 'replication factors' and seem to be the counterparts in virology of the environmental factors, such as low molecular weight nutrients, necessary for bacterial multiplication in host tissues or fluids. Many tissue culture experiments show that 'replication factors' vary from cell type to cell type and are influenced by changes in environment of the cell. In animal infection, variation of availability of 'replication factors' in particular hosts or tissues and under different conditions will affect virus pathogenicity. But few investigations of the influence of such factors comparable in depth to the experiments in tissue culture have been conducted either in animals or organ cultures. Nevertheless, there are signs of this influence in the available studies. There was some parallel between the ability of homogenates of various primate tissues to attach poliovirus and their susceptibilities to virus replication and damage in infection, and some attenuated strains attached to susceptible nerve tissue less strongly than virulent strains. The effect of temperature on virus virulence probably reflects a temperature sensitivity of the enzymes used in virus synthesis rather than an influence on host defence mechanisms. And virus replication *in vivo* can be affected by low molecular weight materials; vaccinia virus infection of mice, like that in cell culture, was enhanced by injection of leucine.

Inhibition Of Host Defence Mechanisms

There has been much work on host defence against viruses but little on the mechanisms whereby viruses overcome this defence.

1. Inhibition Of Humoral Defence Mechanisms

Non-specific humoral defence factors include the low pH of inflammatory exudates and non-specific virus inhibitors in tissues and serum. These inhibitors may be present before infection or be induced by it. The mechanisms of the resistance of viruses to these inhibitors are unknown. Virulent strains of influenza virus appeared to resist such factors in mouse serum more than avirulent strains, but subtle differences in the envelope proteins which might explain the differential resistance have not been investigated.

2. Inhibition Of Cellular Defence Mechanisms

Cellular defence factors include those already present in any tissue the virus attacks, such as possible host nucleases capable of destroying virus nucleic acid, those induced in these tissues, such as the antiviral compound interferon. Then there are those probably present in the phagocytic and other cells of the reticulo-endothelial system. Macrophages definitely ingest and destroy some viruses, but the role of polymorphonuclear leuco-cytes is not as clear. They do not figure so prominently in early inflammatory lesions as they do in bacterial infections.

When a virus does not replicate in any cell, phagocytic or non-phagocytic, this could be due to the absence of one or more 'replication factors' (see above) or to host factors which destroy the virus or interfere with replication. Distinguishing between these two influences in a particular situation is extremely diffi-cult because of the obligate parasitism involved. Nevertheless the distinction has been made occasionally, for example with inter-feron, and this has led (see p.27) to the recognition of possible antagonists of interferon. In seeking to identify virion compo-nents or virus-induced products which inactivate or resist host defence factors, perhaps the first step is to try to distinguish the latter from 'replication factors'. Since this has been done rarely, it is not surprising we know hardly anything about possible virus 'aggressins'.

Virus species and strains within species differ both in the amount of interferon they induce and in their susceptibility to it. Although in many cases virulent strains of viruses induce less interferon or are more resistant to it than attenuated strains, this does not always occur. However, a strict correlation between virulence and induction of or resistance to interferon would not be expected since virulence is almost certainly determined by more than one mechanism. Interferon is produced by the host *in vivo* and it is reasonable to assume that a capacity to reduce its production or resist its action would be an advantage to an invading virus. How could a virus achieve these ends? Early inhibition of host cell RNA and protein synthesis would depress interferon production. Also some viruses appear to produce in tissue culture systems antagonists of interferon; the so-called virus stimulators, virus enhancers and anti-interferons. Whether they are produced in infection and play any role in virus invasion has yet to be assessed.

Macrophages can kill some viruses but not others and, for the latter, ability to resist the killing mechanisms of macrophages and possibly to replicate within them appears to be one of their main virulence mechanisms. Within macrophages, virus is protected from extracellular inhibitors such as antibody; thus, wandering macrophages can spread infection through the blood, lymph and tissues while fixed macrophages can provide an initial focus of infection in larger organs such as the liver. The viral products or mechanisms which determine virus ingestion, survival or replication within macrophages are unknown. Macrophages do not appear to provide a good environment for replication of any virus. Many macrophages in an inoculated population do not become infected; often, viruses survive but do not multiply within macrophages and even when replication occurs yields of infectious virus are small, with much incomplete virus. Within the macrophage the virus surface component may play a role in virus survival by directly interfering with intracellular inhibitors, but equally virus survival may be due to an overall inhibition of macrophage function by a cytotoxic action of the virus. Some viruses such as myxoviruses, vaccinia virus and measles are cytotoxic to macrophages, inhibiting their phagocytic activity towards bacteria. Such cytotoxic effects may be a result of virus replication and thus come under the heading of damage to the host which could aid or hinder, according to the function of the macrophage, further invasion by the same virus or another pathogen. But they would have little relevance to the survival and replication within macrophages of the initial infecting virus particles. On the other hand, components of these initial particles might themselves inhibit macrophage function including viricidal activity and thus contribute to their limited replication. We must know more about the biochemical basis of virus cytotoxicity to decide between these possibilities and recognise the basis of virus survival and replication within macrophages.

3. Inhibition Of The Immune Response

A few days after primary infection a virus must contend with the specific defences of the host. Here, neutralising antibodies supplement the nonspecific inhibitors in body fluids and cellular viricidal mechanisms are strengthened by influence of these antibodies and immune lymphocytes. As regards cellular mechanisms, there is no overwhelming evidence yet that macrophages from immune animals are more viricidal than normal macrophages, although it appears to be true in some cases. However, there are indications that immune lymphocytes are important in defence against some but not all virus infections; they may react with viral antigens and stimulate the infiltration and activity of macrophages.

Viruses could delay or reduce the protective effect of antibody by being 'bad' antigens for inducing antibody, by antigenic variation and by infecting and inhibiting the function of antibody-forming cells. Virus strains vary in their ability to evoke antibody and 'slow viruses' such as the scrapie agent do not appear to induce any. The fact that viruses often have host cell membrane constituents in their envelope proteins provides the possibility of virus antigens being more 'host-like' and therefore 'bad' antigens, but this has not yet been proved. Antigenic variation occurs in influenza and other viruses infecting the respiratory and alimentary tracts and thus must contribute to the ability of these viruses to attack fresh hosts which have neutralising antibodies only against previous variants. There is no evidence however that antigenic variation during the course of infection contributes to virulence, as for example in protozoal diseases. Most virus infections depress but do not shut off antibody synthesis; in some infections it is increased. The way in which antibody-forming cells are inhibited is unknown, but the same mechanisms postulated for inhibition of macrophages

(1) 28

could operate.

Cellular immunity, as judged by graft rejection or delayed hypersensitivity reactions, is depressed in most virus infections and some viruses have been shown to grow in lymphocytes and produce immunosuppression with or without cytotoxic damage. Like macrophage infection, lymphocyte infection provides a ready vehicle for spread of virus infection and is equally unexplained.

Damaging The Host

Overall damage to the host is the culmination of damage to individual cells brought about by virus attack. Here, the ways in which cells can be damaged are discussed.

Virus-induced cell damage may result from a passive role of the virus — a simple repercussion of the process of replication, such as the depletion of cellular components essential for cell life or mechanical harm due to excessive production of virus or its components. Nevertheless, there is increasing evidence that two more positive processes of cell damage occur; namely, virus cytotoxic activity and immunological reaction of the host against virus-infected cells.

There are two levels at which pathologically important cytotoxic activity can operate: biochemical damage without morphological damage and that occurring with morphological damage such as cell lysis, fusion or death. The latter (called here morphological damage) is what is usually meant by cytotoxic (or cytopathic) effect. But both processes must be considered since the former (called here biochemical damage) could cause the decisive pathological damage, for example in nerve cells, even when there is subsequent or accompanying morphological damage in the same or other tissues. In attempts to elucidate these cytotoxic effects, the first question is whether they can be divorced from the process of virus replication and be connected with virus-induced compounds which may or may not be components of the virion. Then we wish to know if the processes of morphological damage can be separated from

(1) 29

aspects of biochemical damage. Finally, we need to know the nature and mode of action of the virus-induced compounds responsible for the cytotoxic effects. Some progress has been made in answering these questions for a few viruses but only in tissue culture experiments. How far the findings can be extended to other viruses and to the pathology of animal infections remains to be seen.

Morphological damage can occur in tissue culture without production of infectious virus. Thus influenza virus and Newcastle disease virus damaged cells which were either incapable or poorly able to support virus replication. Cells were also damaged by poliovirus and vaccinia virus in the presence of chemical inhibitors of virus replication. Furthermore, virulent strains of some viruses such as Newcastle disease virus have greater damaging effects in relation to replication rate than avirulent strains, and the damaging effects of the same virus, such as reovirus, in the same cell line can vary with different cultural conditions which provide similar yields of virus. With regard to biochemical damage, similar experiments have shown that inhibition of host-cell macromolecular synthesis can occur in the absence of production of infectious virus. Finally, it should be noted that pathological damage can occur in animals in the absence of new infectious virus.

Some preformed virion components seem to exert cytotoxic effects. Newcastle disease virus and measles virus produce rapid polykaryocytosis in cell cultures, but only when high virus multiplicities are used; this indicates that preformed products are responsible for the fusion effects. Components of herpes virus also seem to produce giant cells. The penton of adenovirus causes cell rounding and cell detachment from glass. A double stranded RNA from bovine enterovirus causes rapid death of cells without the production of infectious virus. Biochemical damage, more specifically interference with host cell macromolecule synthesis, has been achieved with the fibre antigen of the adenovirus capsid and with a double stranded RNA from poliovirus. Also it should be remembered that large quantities of some viruses such as influenza virus and pox viruses cause rapid toxic effects in animals.

De novo, protein synthesis appears to be responsible for morphological damage of cells infected with poliovirus or vaccinia virus. This was shown by careful time-sequence examinations of the effects on morphological damage of adding and removing compounds which either interfered with the production of infectious virus, such as guanidine, or with protein synthesis, such as puromycin. Similar conclusions, that *de novo* protein synthesis is needed for morphological damage, have been reached for mengovirus, influenza virus and Newcastle disease virus. Also in some instances, inhibition of host cell macromolecule synthesis appears to be due to virus-induced protein synthesis.

Although virus-induced inhibition of host cell macromolecule synthesis could produce decisive biochemical effects in animals and in time will kill cells, in several instances in tissue culture it appears that the rapid morphological damage of cells is not dependent on such inhibition. First, non-infected cells with drug-inhibited macromolecule synthesis were not as damaged as infected cells. Second, sequential observations of virus-infected cells, sometimes coupled with treatment with compounds inhibiting RNA and protein synthesis, showed a lack of parallelism between the appearance of morphological damage and inhibition of macromolecule synthesis. Third, different parts of the capsid of adenovirus have different activities, the penton affecting morphology and the fibre antigen macromolecule synthesis.

In investigating the nature and mode of action of the virus products that are responsible for cytotoxic effects, preformed virion components are the easier target (see above). But the majority of cytotoxic compounds are probably extravirion compounds found in infected cells. Here, the task of identification is more difficult. However, vaccinia specific products from vaccinia virus-infected Hela cells produced cytotoxic effects in fresh uninfected Hela cells. Entry of the virus products was promoted

(1) 31

by magnesium sulphate solution which increased membrane permeability. Obviously there is increasing evidence that viruses produce cytotoxins, but how do they act? Inhibition of host cell macromolecule synthesis or other interference with the functions of the cell could be produced directly by the virus product as diphtheria toxin interferes with protein synthesis. On the other hand, the virus-induced product might act indirectly by releasing autolytic enzymes from the cell's own lysozomes.

Immunopathology is likely to occur in virus diseases because the obligate parasitism involved increases the chances of cell-bound virus antigens occurring and also the existence of auto-immune phenomena. Some viruses incorporate host cell constituents, especially those of membranes, into their structure. Hence antibodies against these virus/host complexes could react with the membrane constituents of infected and normal cells. Also, virus infection may change the host cell membrane con-stituents forming new antigens, the antibodies against which could again react with infected and normal cells. It appears that immunopathology may be involved in some cases of damage in a number of virus diseases such as encephalitis in measles, pox virus rashes, pneumonia from respiratory syncitial virus, yellow fever, haemorrhagic dengue, mumps, and coxsackie-B virus infection. In these diseases the evidence is mostly suggestive. But for lymphochoriomeningitis in mice and aleutian disease in mink we have the virus counterparts of tuberculosis, where suffi-cient solid experimental evidence has accumulated for us to be reasonably sure that immunopathology plays a major role in the observed damage.

Immunopathology is such an attractive explanation for virus damage that it is receiving much current attention. Perhaps a few words of warning against too easy assumptions of its com-plicity in cases of virus damage may not be out of place. First, mere demonstration that an infected host is immunologically sensitive to virus products by a diagnostic test such as a skin test is no proof of implication of sensitivity in the main pathological effects of disease. Second, the lack of knowledge of the mechanisms of direct virus cytotoxicity adds to the difficulty of distinguishing such mechanisms from immunopathological ones. Third, it should be remembered that, although interesting, the number of immunopathological cases are probably small compared with those due to direct virus damage.

Tissue And Host Specificity In Virus Infections

Certain viruses attack certain tissues (e.g., poliovirus, enteric and anterior horn tissue) or hosts (e.g., variola, primates; foot and mouth disease virus, domestic animals) but not others. In fact tissue and host specificities are two of the most documented aspects of virus disease. On the other hand, studies of the biochemical basis of these phenomena in virus infections are even more in their infancy than similar studies in bacteriology.

The real difficulty in studies of host and tissue specificity is not lack of ideas of possible explanations for the phenomena but design of experimental systems to investigate them in a manner relevant to natural infection. Clearly, the availability of 'replication factors' in host cells and their surrounding fluids and their variation under different environmental conditions probably determine many cases of host and tissue specificities. Similarly, other cases will depend on variation in levels of antiviral substances from host to host and tissue to tissue, or differential induction of interferon and immune mechanisms either in level or in time. Also, the route of infection may play some role in tissue specificity. One susceptible tissue may be easily accessible to the incoming virus and become infected, whereas another equally susceptible tissue may be protected by a barrier of cells which either do not support virus replication or destroy the virus. For example, in mice the Kuppfer cells of the liver seem to protect the parenchyma cells from infection with blood-borne influenza and myxoma viruses; and infection via the bile duct circumvents the barrier. Methods of investigating the possible reasons for tissue and host specificity are discussed below.

The first essential is that the specificities of animal infection should be retained in the experimental system. Since virus susceptibilities change when cells differentiate in normal tissue cultures, the latter cannot be used directly to investigate host and tissue specificity in natural infection. Nevertheless, studies of differing cell susceptibilities in such cultures might serve as models for adaptation to the systems described below.

Short-term studies with primary cell cultures or suspensions of relevant tissues have yielded most of our available information on tissue and host specificity. Using such preparations and membrane fractions from them, Holland and his colleagues provided good evidence that the presence or absence of surface receptors for poliovirus determined susceptibility or resistance to infection of different primate tissues and of primate and non-primate tissues; in particular, cell resistance to infection disappeared when the receptors were by-passed by using virion RNA as the infecting material. In other cases, experiments with primary cell cultures have shown that cell receptors determining initial adsorption are probably not the important factors in susceptibility. The MHV (PRI) strain of mouse hepatitis virus infects PRI mice but not C_3H mice and this difference in host specificity is reflected in the susceptibility of liver macrophages. Yet adsorption of MHV (PRI) virus to resistant mouse (C_3H) macrophages was similar to that occurring with susceptible mouse (PRI) macrophages; penetration of the resistant macrophages seems to have occurred, but there appeared to be no uncoating. Similarly, work with primary cell cultures showed initial adsorption but absence of penetration occurred with insusceptible cells for Rous sarcoma virus and feline herpes virus.

In the future, experiments with organ cultures coupled with those in animals may prove of equal importance to work with primary cell cultures in studies of host and tissue specificity. Unlike tissue cultures, organ cultures usually retain their parent specificities of natural infection and when this is not so, it indicates that the specific or non-specific host defence mechanisms present in animals but absent in organ cultures may play an important role in the tissue or host specificity. Using organ cultures and whole animal experiments (Toms, Rosztoczy & Smith, 1974; Rosztoczy, Toms & Smith, 1973), some of the mechanisms of the tissue specificity of influenza virus in ferrets have been elucidated. The virus proved more ubiquitous than expected, infecting bladder, uterus (also human endometrium), oviduct and conjunctiva as well as the respiratory tissues, nasal mucosa, trachea and lung. Furthermore, determinations of minimal infective dose in organ culture indicated that bladder and uterus were almost as easily infected as the most susceptible respiratory tissue, nasal, mucosa, and significantly more so than trachea and lung. Although in animal experiments the urogenital tissues could be infected by local inoculation they failed to be infected by either nasal or blood inoculation of large quantities of virus. This indicates that one factor operating in localisation of the virus in the respiratory tract is a barrier phenomenon preventing blood-borne virus infecting other susceptible tissues. Despite the clear-cut differences in ease of infection of the various susceptible tissues indicated by the minimal infective dose estimations, observations on virus replication kinetics over one cycle showed that for all susceptible tissues new virus appeared at the same time, 5-6 hours after infection. Thus differences in susceptibility of respiratory tract tissues (and urogenital tissues) are probably not due to intracellular happenings in the individual susceptible cells. Differences either in number or availability of individual susceptible cells in the various tissues may be the explanation. Alternatively, the amount of virus that enters or is released from the susceptible cells may vary in the different tissues.

PATHOGENICITY OF MYCOPLASMAS, FUNGI
AND PROTOZOA

The pathogenicity of mycoplasmas, fungi and protozoa has not been investigated in any depth, but there are signs that these microbes will receive more attention in the near future.

Mycoplasmas

Mycoplasmas (micro-organisms without a rigid cell wall) such as *Mycoplasma pneumoniae* cause respiratory and urogenital tract infections in man and animals. Identification of virulence markers and determinants has not occurred because estimations of relative virulence of mycoplasma strains are only just beginning and like those of viruses are hampered by inaccuracy of viable counts. The host defence mechanisms which may operate against mycoplasmas on mucous surfaces are unknown. However, they may be adversely affected by products from commensals because of their extreme sensitivity to lysis by many factors. How such lytic factors are overcome, and the mechanisms of adherence of mycoplasmas to mucous surfaces are unknown.

Nothing is known of factors determining multiplication *in vivo*. Materials lethal for mycoplasmas are found in tissue extracts and may include lysolecithin. Phagocytosis and intra-cellular killing of mycoplasmas by polymorphonuclear and mononuclear cells has been observed *in vitro* and *in vivo*. But whether mycoplasmas produce aggressins which interfere with these host defences is not clear.

Mycoplasmas may damage host tissues by mechanisms similar to those of bacteria and viruses, passively by excessive usage of nutrients such as arginine, or directly, by the production of hydrogen peroxide or neurotoxins such as those of *Mycoplasma gallisepticum* and *Mycoplasma neutrolyticum*. Also, close adherence of mycoplasmas to cells may change surface antigens leading to hypersensitivity or auto-immune effects. Nothing is known of the basis for tissue and host specificity of mycoplasmas.

Fungi

Many fungi produce animal disease such as thrush, dermatitis, 'farmers lung', and mycotic abortion. Some have mycelial (arthrospore) and yeast forms differing in virulence; they seem particularly appropriate for comparative studies which might reveal the determinants of virulence. Quantitative virulence comparison of yeast and arthrospore forms should be relatively easy, since both forms are easily counted *in vitro*.

Antifungal mechanisms operative on mucous and other body surfaces include inhibitory activity of bacterial commensals (since antibiotic treatment can result in fungal infection) and fungistatic materials in various secretions such as those of the conjunctiva and saliva and the fatty acids in teat secretions of domestic animals. The factors which determine the survival of fungi on mucous membranes are not known, nor are those which determine replication in animal tissues. Within the tissues, humoral antifungal factors can be detected and killing by phagocytosis also appears important in resistance to some mycoses. Many fungi are larger than phagocytes and mere size may prevent ingestion, thus rapid growth alone can be an aggressive mechanism. A capsular polysaccharide prevents the phagocytic ingestion of virulent strains of *Cryptococcus neoformans* and, like brucellae and tubercle bacilli, *Candida albicans* and *Histoplasma capsulatum* can survive and grow within phagocytes. Inhibition of intracellular killing may be due to fungal aggressins comparable to those produced by brucellae. Production of aggressins may be related to yeast forms; the non-invasive mycelial dermatophytes probably lack the powerful aggressins of the yeast-like fungi which cause deeper mycoses.

In some mycotic diseases, mechanical blockage by large mycelia probably damages the host. Fungi produce in foodstuffs powerful toxins such as aflatoxin and sporidesmin. The chemical constitutions of these toxins are known and their action in various hosts have been studied. Nevertheless at present they are comparable to botulinum toxin, i.e., poisons produced exterior to the host. They may well be produced in mycotic infections but this has yet to be demonstrated. Peptidases, collagenase and elastase appear to be involved in dermatophyte damage, but hypersensitivity to ill-defined glycopeptide cell-wall products probably explains to a large degree the pathology of most fungal skin diseases. Whether or not hypersensitivity enters into the main pathology of oral or deeper mycoses is largely a matter for speculation.

An example of tissue specificity in fungal infections, namely the growth of *Aspergillus fumigatus* in placental tissue, which causes mycotic abortion in ewes and cattle, may have a nutritional basis. A material which stimulates spore germination is concentrated in bovine foetal placenta but its nature is yet unknown (White and Smith, 1973). Also, the lack of sebaceous glands which excrete mycostatic fatty acids may allow growth of *Trichophyton* spp. in certain areas, for example, between the toes.

Protozoa

The microbial factors responsible for the pathological effects of protozoal infections have received little attention. Quantitative comparisons of virulence present no insuperable difficulties. The major difficulty in identifying virulence factors appears to lie in the versatility of protozoa. The different phases of their life cycles, their different morphological forms and their antigenic plasticity all have profound influences on pathogenicity. Keeping one strain in one form for repeated experiments on a particular problem in pathogenicity is a major operation.

Few detailed studies appear to exist on the influence of

membrane environment and commensals on the ability of protozoa to survive on and penetrate mucous surfaces, although some gut bacteria seem to be an essential food for *Entamoeba histolyticum*. Nutrition seems important in protozoal pathogenicity. Malarial attacks seem to be stimulated by p-aminobenzoic acid and methionine and the parasite seems to grow within red cells because it needs haemaglobin to form haemazoin. Host defence mechanisms against protozoal attack include humoral factors (antibody and complement) in some infections (e.g., trypanosomiasis) and phagocytosis in others (e.g., malaria and leishmaniasis). It has been suggested that interferon (page 26) may contribute to defence against malaria. The mechanisms whereby protozoa avoid or inhibit host defence mechanisms are obscure. Trypanosomes appear to avoid lysis by antibody and complement, and malarial parasites avoid destruction by macrophages by changing their antigens *in vivo* as infection proceeds. The high mobility of some protozoa may prevent phagocytosis and entamoebae appear to produce factors which kill leucocytes. But the author is not aware of a protozoal product which prevents ingestion by phagocytes. *Toxoplasma gondii* produces a factor which promotes penetration of host cells and, as for bacterial aggressins, enhances its virulence for mice. Since *Leishmania* spp. survive and grow within phagocytes they may produce as yet unidentified aggressins similar to those of brucellae.

The overall toxic effects of protozoa have been investigated, especially those of *Plasmodium* spp. Clearly, cell destruction by intracellular growth of protozoa could contribute to damage in such diseases as malaria and leishmania. A microbial toxin has not yet been recognised as being unequivocally responsible for the main pathological effects of any protozoal disease. In malaria, increase of capillary permeability leading to shock and brain damage appear to be mediated by kallikrein, kinins and adenosine but a malarial toxin responsible for the release of such host products has not been demonstrated. The lytic effects of *Entamoeba* spp. are known but the mechanisms of cell destruction are not clear; lysosomal enzymes may be transferred from the entamoebae to the host cells by tubules formed between the two cells on contact. Toxic products of trypanosomes and toxoplasmas are known, but their importance in disease is not clear. Hypersensitivity and auto-allergic phenomena occur in protozoal infections, but their responsibility for important pathological effects is hard to judge. In malaria, antibodies to host antigens changed by red blood cell parasitisation may, by opsonisation, promote phagocytosis and destruction of unparasitised red blood cells.

Tissue and host specificity occurs in protozoal infection. The connection between growth in red cells and the requirement of the malarial parasite to form haemazoin has been mentioned. It has been shown (Butcher, Mitchell and Cohen, 1973) that host-specificity for *Plasmodium knowlesi* infection is not correlated to the antiprotozoal or other properties of the different sera but with ability of the merozoites to adhere to the different red blood cells. The nature of the red blood cell receptors of the susceptible species (humans and three monkey species) is unknown.

BOOKS AND ARTICLES FOR REFERENCE

Ajl, S.J., Kadis, S. and Montie, T.C. (1970), in *Microbial Toxins*. London and New York, Academic Press.

Dubos, R.J. and Hirsch, J.G. (1965), in *Bacterial and Mycotic Infections of Man*, 4th ed. Philadelphia, Lippincott.

Dunlop, R.H. and Moon, H.W. (1970), in *Resistance to Infectious Disease*. Saskatoon, Modern Press.

Howie, J.W. and O'Hea, A.J. (1955), in *Mechanisms of Microbial Pathogenicity*. Cambridge, Cambridge University Press.

Mims, C.A. (1964), Aspects of the pathogenesis of virus disease. *Bact.Rev.*, **28**, 30.

Smith, H. (1960), The biochemical response to bacterial injury. *In Biochemical Response to Injury*. p.341. Oxford and Edinburgh, Blackwell.

Smith, H. (1968), The biochemical challenge of microbial pathogenicity. *Bact.Rev.*, **32**, 164.

Smith, H. (1972), Mechanisms of virus pathogenicity. *Bact.Rev.*, **36**, 291.

Smith, H. (1973), Microbial interference with host defence mechanisms. *In Monographs in Allergy*, **9**, 13-38.

Smith, H. and Pearce, J.H. (1972), in *Microbial Pathogenicity in Man and Animals*. Cambridge, Cambridge University Press.

Smith, H. and Taylor, J. (1964), in *Microbial Behaviour in vivo and in vitro*. Cambridge, Cambridge University Press.

Smith, W. (1963), in *Mechanisms of Virus infection*. London and New York, Academic Press.

Tamm, I. and Horsfall, F.L. (1965), in *Viral and Rickettsial Infections of Man*, 4th ed. Philadelphia, Lippincott.

Wilson, G.S. and Miles, A.A. (1964), in *Topley and Wilson's Principles of Bacteriology and Immunity*, 5th ed. London, Edward Arnold.

RECENT REFERENCES

Butcher, G.A., Mitchell, G.H. and Cohen, S. (1973), *Nature (Lond.)*, **244**, 40.

Punsalang, A.P. and Sawyer, W.D. (1973), *Inf. Imm.*, **8**, 255.

Rosztoczy, I., Toms, G.L. and Smith, H. (1973), *Lancet*, **1**, 327.

Toms, G.L., Rosztoczy, I. and Smith, H. (1974), *Brit.J.exp.Pathol.*, **55**, 116.

Uchida, T., Pappenheimer, A.J. Jr. and Harper, A.V. (1972), *Science*, **175**, 901.

White, L.O. and Smith, H. (1974), *J.med.Microbiol.*, **7**, 27.